MACHINE LEARNING

The Ultimate Beginner's Guide to Machine Learning

Edward Mize

First published in 2019 by Venture Ink Publishing

Copyright © Teaching Nerds 2019

All rights reserved.

For more information about the contents of this book or questions to the author, please contact Edward Mize at edward@teachingnerds.com

Disclaimer

This book provides wellness management information in an informative and educational manner only, with information that is general in nature and that is not specific to you, the reader. The contents of this book are intended to assist you and other readers in your personal wellness efforts. Consult your physician regarding the applicability of any information provided in this book to you.

Nothing in this book should be construed as personal advice or diagnosis, and must not be used in this manner. The information provided about conditions is general in nature. This information does not cover all possible uses, actions, precautions, side-effects, or interactions of medicines, or medical procedures. The information in this book should not be considered as complete and does not cover all diseases, ailments, physical conditions, or their treatment.

You should consult with your physician before beginning any exercise, weight loss, or health care program. This book should not be used in place of a call or visit to a competent health-care professional. You should consult a health care professional before adopting any of the suggestions in this book or before drawing inferences from it.

Any decision regarding treatment and medication for your condition should be made with the advice and consultation of a qualified health care professional. If you have, or suspect you have, a health-care problem, then you should immediately contact a qualified health care professional for treatment.

No Warranties: The author and publisher don't guarantee or warrant the quality, accuracy, completeness, timeliness, appropriateness or suitability of the information in this book, or of any product or services referenced in this book.

The information in this book is provided on an "as is" basis and the author and publisher make no representations or warranties of any kind with respect to this information. This book may contain inaccuracies, typographical errors, or other errors.

Table of Contents

Who is this book for?

Data comes in all forms, including texts, numbers, images, tables, videos, transactions, and more. In fact, everywhere we go, we become "datafied", so they say.

Our smartphones can track our locations, and that itself is data. Whenever you browse the web, you leave a data trail. Your every interaction in social networks, every single one of them is data.

For this reason, privacy is highly important in Data Science. To people who want to learn how they can use data for various purposes, machine learning will be an interesting subject.

You can do many things with data. These are reasons why business-leaders are quite eager to hire people with talent and experience in this field.

This book will introduce you to the breakthrough of what machine learning is.

It takes supercomputers with the ability to handle massive volumes of data and tremendously quick learning ability to make superfast decision-making capability based on that learning. Today's generation is all about this.

Even smartphones today have huge thinking capacities. Imagine what supercomputers can do.

Businesses, weather forecast stations, social media companies, streaming services, and more can are all using machine learning. With so many areas you can bring it in, it makes sense many people want to master it.

What will this book teach you?

Many people commonly misunderstand machine learning. To those who are not acquainted with the term, machine learning is about making a computer smart enough that it is able to learn from new data.

After some time, the computer learns to predict further data based on what it's been fed. But how, you ask?

The concept is at its core is the same as a simple algebraic problem. For instance, you have X and Y values. If you have the value of X, finding out Y is pretty much a piece of cake.

Machines can learn the same way as when you solve Y given the value of X. A computer builds a curve or modifies it whenever some data is fed to it.

Say you are calculation X and Y using the computer by feeding it new information. A computer capable of learning can give you the value of Y the next time you give a new value for X.

Unlike when you were at a school where you need to physically draw a curve, the computer it by generating the curve's equation.

For years now, researchers have been building algorithms on machine learning. They understand how practical it is to have a machine capable of learning.

As a result, their techs have successfully evolved now to a stage where businesses can use machine learning as a concrete tool if you have an idea how.

You can use machine learning in many ways, one of which is by helping businesses make pointedly better decisions.

However, this field is rather neglected and if someone in your office can remotely understand it, it is probably accidental, not intentional. The business field is only one of the areas that machine learning finds use.

In this book, we will try to help you understand what machine learning is and what its various applications are.

If you aspiring to be a data scientist, a decision scientist, or just someone whose field involves mining tons of data, this book serves a good read to get a grasp of the topic. This book will serve as your guide to learning the essence of machine learning.

Introduction

There is data everywhere. Did you know that the last two years alone created 90% of the data in the world? The rough estimation of our current output of data is around 2.5 quintillion bytes per day.

This number is only set to grow in the coming years together with more of the world becoming connected to the increasing number of electronic devices.

Since 2016, the Internet population grew again by 7.5 percent that includes more than 3.7 billion humans and this is only the latest data on 2017.

This year, you can expect that this number grows only more. In terms of Internet data, the U.S. alone consumes around 2,700,000 gigabytes of internet data and that's within every minute.

Every day, Google processes about 24 petabytes of data. In every hour, Facebook uploads around 10 million photos. In Twitter, there are about 400 tweets a day.

Within ever second, someone uploads an hour worth of video. It is not only Internet-based media services and platforms that are growing but others as well.

These are only a fraction of all data surrounding us, processed every second of every day. With the fact that we leave data in our every wake, it is not surprising that there is a lot of buzz in machine learning. Can machines really learn? Can you really predict the future and make better decisions using big data?

Machine Learning in the Real World

There are many forms of machine learning in the real world. For instance, your movie recommendations in Netflix – those are an example of machine learning.

Another example includes Facebook's capability of spotting the faces of your friends, your Amazon product recommendations or the potential date matches on your dating app.

One of the classic examples that show you the capability of machine learning is the self-driving car. The truth is no one is actually capable of creating a car that can drive on its own. There is no programming that can do that.

This is why a self-driving car cannot actually drive itself. In fact, even if you know how to drive, you cannot explain how to yourself or to other people.

For a self-driving car to learn how to drive, it has to learn by experience. You need to drive it first many times while the car also observes how other people drive.

Without observing, experiencing and practicing it, a self-driving car will not be able to learn how to drive. These are the keys how machines learn, change and adapt.

They are able to do this by using algorithms designed to learn through experience. Now, all humans simply need to do is create an algorithm for things they are unable to make a program.

To many computer scientists, machine learning is a bigger innovation than electricity, personal computers, and the Internet itself.

With machine learning, we have now entered the second stage of the information age. The first stage involves writing programs computers and the second stage, which we are currently in, involves computers that can program themselves simply by looking at data.

This is why in the years to come, businesses will have one thing in common – machine learning.

The following chapters of this book introduce you to essential concepts, applications, essence, and importance of machine learning, data mining, and statistical pattern recognition.

You can learn about how it is an interdisciplinary field. Furthermore, you can learn how to distinguish statistics and machine learning

Chapter 1

What is Machine Learning?

Learning is a process that covers a wide range of other processes. This makes giving it a clear definition a little difficult.

According to the dictionary, learning is an activity where you gain knowledge or skill through studying, being taught, practicing or by experience. Others say that learning happens when your behavioral tendency is modified by experience,

We are all interested in finding out about how both humans and animals learn, which is why we have psychologists and zoologists.

This book tackles the same thing that zoologists and psychologists want to know – how to "learn." Specifically, this book will help you understand machine learning and how you can use data to predict the future.

Unlike machines before, which only know the things included in its programming, modern machines are different.

Today, we are in the center of a historic moment where computers are given the ability to learn from experience.

Modern machines can now adapt or change to new data or experience. This is the so-called "machine learning."

Use Data to Predict the Future and Make Good Decisions

Machine learning can do more than just learn new skills and knowledge or modify its behavior by experience. One of the most important applications of machine learning is in making a good decision.

With help from previous data fed into a computer, you can learn the possible outcomes or predict the future that allows you to make better choices.

There are many people thinking it takes meeting your goals and getting ahead of your competition to say you made a good decision.

However, a good decision has more to it than that. Decisions affect everyone, the whole world around us. Just one wrong decision you make can have a negative impact on everything around you.

This makes it important for an individual to have the skills in making good decisions. Machine learning is one of the more advanced methods of structuring and making decisions based on data.

This innovation is indescribably essential if you want to understand what direction technology and society goes is heading right now.

How Does a Machine Learn?

In machine learning, a computer can do the natural way that humans and animals learn from experience.

A computer uses machine learning algorithms, which in turn utilizes computational methods to learn directly from data without the need to use a model, such as a predetermined equation.

The machine learns to adapt from samples

Instead, the algorithms' performance improves as it adapts to the number of samples it has access to increase learning. This allows a machine to learn from experience and examples even without explicit programming.

With machine learning, there is no need to write a code. Instead, you use a generic algorithm, which you feed new data.

The machine builds logic to make decisions

From the new data that you fed the algorithm, the machine builds logic to make decisions, predict the next thing that will happen, and do a lot of other things.

One example of an algorithm used in machine learning is the classification algorithm. In this system, data is put into different groups. This algorithm is often used in statistical learning and applied data mining.

The machine learns every time it changes data, program, and structure

With regards to machines, it can be said in general that machines can learn every time it changes its data, program or structure.

The changes are based on new data inputs or as a response to the external information. The machine's changes happen in such a way that its expected future performance is improved.

Some of the changes that occur are not necessarily understood as something that can be called learning because and often fall within other discipline's areas.

However, we can be satisfied and say that a machine has learned when, for example, a speech-recognition machine hears a number of samples of a person's speech and it improved.

Machine learning leads to changes in systems performing AI-related tasks

Typically, machine learning points to the changes in structures and systems that perform AI-related tasks. These tasks comprise of recognition, diagnosis, design, prediction, robot control and others.

The changes to the machine could be improvements to the existing performing structures. It could also be an initial synthesis of new structures or systems.

More Data and More Questions Equals More Answers

Through machine learning, your questions can have better answers the more data you have.

Algorithms look for natural patterns in data, which generate insight that help you find better answers to your questions, make better decisions and make more accurate predictions.

Every day, people from different fields use machine learning in making critical decisions.

People use them in stock trading, medical diagnosis, weather prediction, energy load forecasting, and more. Even media sites use machine learning to filter through the millions of Internet data.

This allows them to give you recommendations on songs, movies, books, products, and more. Big data made machine learning particularly important when solving problems in

- *Medical Diagnosis,* in diagnosing if a patient is a sufferer or non-sufferer of a particular disease.

- *Face Detection*, in identifying faces in images or in indicating if a face is present within the image.

- *Weather Prediction*, for instance, in predicting whether it will or will not rain tomorrow.

- ***Email Filtering***, in classifying emails into spam or not-spam.

Some interesting real-world examples where you can use a classification algorithm include classification of email spams, handwritten digit recognition, speed recognition, image segmentation, and DNA sequence classification.

If you are looking to classify certain things, you can use this machine learning classification algorithm to sort your data.

Other real-world applications of machine learning include motion detection and object detection in the field of computer vision and image processing.

Machine learning is also used in algorithmic trading and credit scoring in computational finance, drug discovery, and tumor detection in computational biology, among others.

Chapter 2

The Need for Machine Learning

Artificial Intelligence (AI) gave rise to the field of machine learning. We use AI because we wanted to create better and smarter machines.

However, we find it is hard to program more complex challenges that constantly evolve. We are unable to program anything except for simple tasks like finding the shortest path to get from point A to point B.

It made it clear that the only way to program more complex tasks is by teaching the machine how to do it, specifically, to let it learn from itself.

However, a computer that learns from itself is a lot like a child that learns from itself. This gave rise to the development of machine learning as computer's new capability.

The human is the only one that can find patterns in data, but it cannot process a large amount of data within a minimum time

Nowadays, machine learning is just about everywhere. It is around many pieces of technology that most of us do not realize when we use it. Only the human brain can find patterns in data.

However, the human brain is unable to process large data in a short amount of time. This is where machine learning comes into play.

When data is massive, the time needed to compute is also increased. Machine learning comes into play here to help people in processing such a large data in the shortest possible time.

Just like cloud computing and big data, machine learning is gaining importance because of its contributions in helping evaluate big pieces of data and helping the task of data scientists.

Several factors drive the demand for machine learning – Cloud Computing and Big Data

The need for machine learning is everywhere. You must be wondering, what is driving this huge demand for computers capable of learning?

There are several factors that influence the demand, one of which is the fact that cheap machine power is now readily available. In particular, cloud continuous to make computation cheaper.

The more that computation becomes cheaper, the more that the impossible becomes possible. With more things becoming possible, the more people start doing them.

In the end, you meet people whose work relates to making the impossible possible. This is when you meet people like data scientists, data analysts, software engineer, and more.

Another factor that drives the need and demand for machine learning is the last few years' explosive growths of data.

In the last few years, there has been a big buzz around Big Data, particularly on collecting, storing and doing a calculation with it. On the other hand, there is less discussion on the fact that you can obtain useful insights from it.

Machine learning, as it turns out, is particularly suited to this. Companies with big data initiative are beginning to realize this, which serves as a big demand for machine learning.

Big data and machine learning are not only about improving the bottom line of the company. The most important thing is survival.

New, Better Algorithms vs. Existing Algorithms

Today, it is common for startups to use a data-centric approach to upset the present markets. There is the risk of extinction for fixed companies that do not understand this trend.

For companies who do not yet understand machine learning, they wonder what work they need to do concerning its growing demand. Is there the need for new and better algorithm?

The answer is most likely yes. This is because there is the possibility that someone will create a new algorithm that can change everything, which could not really be a bad thing.

Nevertheless, there might not be the need for that since there are already many good algorithms available today to solve many problems readily available.

There are probably hundreds of classification algorithms that exist today to solve a real-world problem. There is a paper about this that evaluated more than a hundred qualifiers used in wide array of datasets.

There are hundreds of them you can use for a classification problem, but the assessment showed that the best classifier was RDF, which was not exactly new.

This shows that you might not exactly need to create a new algorithm. Of course, the need for newer algorithms will never go away. This is necessary for the advancement of machine learning.

However, it is a fact that right now there is a huge of amount of work needed to be done, which needs existing algorithms to be moved to the practical world.

Perhaps it is most important right now to make existing algorithms more powerful and usable. This is more important than building the perfect model algorithm that is utterly elusive.

The reality for most projects is that the perfect model is not going to be the final product. Most often, the model will only be a part of the complete application.

Machine Learning Saves Time and Money

The time and money savings accompanying machine learning are two of the driving factors behind it. These two advantages have the potential to have a radical impact on an organization's future.

Machine learning has a particularly tremendous impact in the customer care industry. Through machines, customer care personnel can do things more efficiently and quickly.

The importance of machine learning is not only found in customer service but also in virtual assistant solutions.

Through machine learning, tasks that would usually need a live agent to perform can now be simply automated. With that, time can be freed to focus more on important tasks that a human can perform best.

With machines, a valuable agent can invest his or her time for complex decision-making that a machine cannot handle easily.

The process can even be further improved where the decision of whether to let a human or machine handle a request is eliminated. This can happen through a machine's adaptive understanding that allows it to learn of its limitations.

Being aware of its limitations, it can let the request go to humans when it does not have the confidence to provide the exact solution. In the last few years, machine learning has successfully made significant improvements.

However, it is still too far from reaching the same human performance when there are times it still needs human assistance to complete its task.

Machine learning is developed for complex programs humans cannot code directly

Some of you are probably wondering why machines have to learn or why not design one that can perform as you desire right from the start.

As mentioned before, machine learning is developed for tasks that are programs that are too complex for a human to code directly.

There are tasks that are too complex in which coding them is not only highly impractical but also probably impossible.

For these complex tasks, humans might not be able to work out all the code and nuances explicitly for the program to work. Instead, you use a machine learning algorithm.

By providing a massive amount of data and letting the algorithm do its work and explore the data, the machine can search for a model that can accomplish what the programmer set it out to complete. This is only one of the several reasons that make machine learning important.

Understanding how machines learn can help us understand how humans and animals learn as well

Another one of them is the fact that the achievement of machine learning can help us better understand how humans and animals learn.

The way that humans, animals, and machines learn has several parallels between them. Many of machine learning techniques are derived from psychologists' efforts in understanding human and animal learning.

Psychologists try to make their theories about how human and animal learning more accurate by using computational models.

Machine learning uses the same models to create systems can learn from themselves. It also seems likely that the systems machine learning researchers use can elucidate some characteristics of biological learning.

Machine learning has practical engineering significance

Besides those mentioned above, there are several significant engineering reasons that make machine learning important. For instance, some engineering tasks can only be defined well with the use of examples. It could be possible to identify the input/output pairs but not specify the succinct connection between the inputs and the preferred outputs.

Machine learning can create adaptable internal systems and structures

In engineering, what we would like is for the machines to have the ability to modify their internal systems and structures to develop the correct outputs for the inputs.

Therefore, you can appropriately compel the systems/structures' input/output function to estimate the examples' implied relationship. This is exactly what machine learning can do.

Machine learning can improve existing machine designs

Machine designers commonly produce a machine that does not only work well but is not even desired in the environments they are used.

This could be because the characteristics of the working environment are not completely known during the machine's design time.

To counter this, engineers can use machine learning to redesign and improve existing machine designs.

Explicit Coding of a Large Amount of Data is Impossible for Humans

You have heard before. Machine learning is developed because it would be completely impossible for humans to explicitly code a programming with an extremely large amount of data.

Through machine learning, humans instead designed algorithms that allow machines to learn the knowledge, so it can learn to do what humans cannot.

Machine learning helps machines to adapt to changing environments

It is natural for an environment to change as time goes. Through machine learning, machines like computers gain the ability to adapt to the changing environments around them.

This is a vital characteristic of machine learning as it would mean there would no longer be the need for constant redesign. With this, even machine can evolve.

Machine learning can extract important relationships

There is a high possibility that within large amounts of data, important correlations and relationships are hidden. These relationships are not easy to find and extract, which is what machine learning is for.

You can usually use the methods used in machine learning to withdraw these relationships. This is what you call data mining.

Machine learning might be able to track constant changes due to new knowledge

Every hour of every day, humans discover new knowledge about tasks. For example, new words are invented, and so vocabulary always changes.

Around the world, there is a constant stream of events that change an aspect of the society. It is impractical to continuously redesign AI systems to follow these changes. However, a computer's machine learning might be.

Chapter 3

Machine Learning – A Look into the Past

Machine learning converges from a number of sources or foundations. Several different disciplines or traditions bring in various methods and vocabulary integrating into a more amalgamated discipline.

To better understand machine learning, it is best to become familiar with the separate disciplines that contributed to it.

History of Machine Learning

Machine learning is a scientific endeavor that started and grew out of human's pursuit of artificial intelligence. In the early days of AI research, some investigators were fascinated in having machines that can learn from data. In approaching the problem, they used several symbolic methods along with "neural networks."

Neural networks were computer systems models on the human brain and nervous systems. At that time, most neural networks were perceptrons and other models, which were actually the generalized linear model's reinventions.

In addition to neural networks, researchers also used probabilistic reasoning, which is used for automatic medical diagnosis.

Artificial Intelligence and Machine Learning Split

Artificial intelligence and machine learning split from each other when the prominence on the logical, knowledge-based approach increases.

Probabilistic systems were simply too inundated by both theoretical and practical problems in terms of data acquisition and representation. This caused more problems between AI and machine learning.

Statistics is No Longer Involved in AI

In 1980, artificial intelligence no longer favors statistics. By that time, expert systems (computer systems that imitates a human expert's decision-making ability) dominates the field of artificial intelligence.

The symbolic/knowledge-based learning approach continues still within AI, which led to inductive logic programming.

However, there was no longer a more statistical approach in AI.

Statistics went to two branches of machine learning – *pattern recognition* with focus on recognizing regularities and patterns in data and *information retrieval (IR)*, an activity that focuses on obtaining information resources significant to a collection of information resources' information need.

AI and CS Abandoned Neural Networks Research

Similarly, both computer science and AI abandoned neural networks research at around the same time, which was continued outside of this field.

It went to become "connectionism", an approach in the fields of cognitive science, artificial intelligence, and philosophy of mind.

Researchers such as Hopfield focused on the discipline gaining prominent success in the mid-1990s.

Machine Learning as a Separate Field

In the 1990s, machine learning was restructured and started to flourish as a separate field. It changed its goal – instead of achieving artificial intelligence, it now aims to tackle more solvable problems with real-world nature.

Similar to AI, machine learning let go of the symbolic approach it inherited from the former.

Its focus, instead, shifted to the use of models and methods adopted from probability theory and statistics.

With the increased availability of digital information and the Internet, machine learning also benefits from access to Big Data. This has led to ML and data mining to sometimes overlap and use the same methods.

Machine Learning vs. Data Mining

You can roughly differentiate machine learning and data mining as the former's focus is prediction, while the latter is on discovery.

Machine learning focuses on making a prediction based on identified properties that the program learned from the training data.

On the other hand, data mining's focus is on the discovery of data properties that were previously unknown.

In the Knowledge Discovery brand of data mining, the discovery of formerly unknown data properties is within the analysis step.

Machine learning and data mining often intersect in many ways, where the latter uses ML methods but with a somewhat different goal. Similarly, machine learning uses data mining methods, such as unsupervised learning.

Many people get confused between the two because of their basic assumption of what machine learning and data mining work with.

In Knowledge Discovery and Data Mining (KDD), the main task is to discover formerly unknown knowledge. Machine learning, on the other hand, the performance is evaluated in connection with the ability to imitate known knowledge.

In machine learning, a supervised method can easily outperform unsupervised method when evaluated in reference to known information.

On the other hand, usual KDD tasks cannot use supervised methods because there are no training data available. Another of the difference between the two is the ties of machine learning to optimization.

Disciplines That Influenced Machine Learning

- ## Statistics

    Finding out how you can best use the drawn samples from unknown probability distributions to help choose which distribution the new sample will be drawn is one of the most enduring problems in statistics.

    There are similar problems, which uses statistical methods that can be considered as instances of machine learning. This book will also explore those methods.

    Statistics and machine learning are closely interconnected fields.

    The ideas and approaches of machine learning, such as methodological principles and theoretical tools, have had a long connection with statistics.

Similarly, some statisticians adopted machine learning methods that lead to a combined field known as "statistical learning."

- ## Bayesian Methods

  Bayesian methods are used in a broad range of fields including machine learning. Bayes' Theorem is the basis of the Bayesian variety, a branch of Machine Learning.

  It also serves as the basis for computing the probabilities of hypotheses. The Naïve Bayes is a simple but powerful algorithm used for predictive modeling utilizing the Bayes' Rule.

- ## Brain Models

  Artificial systems do not necessarily need to copy nature. However, both AI and machine learning successfully and convincingly draw inspirations from both neuroscience and psychology in leading significant innovations.

  One very prominent example of this is the deep neural networks. Important machine learning techniques are based on systems of nonlinear elements.

  Recently, an IBM research team also took inspiration from the brain and developed computational models of memory and attention.

The goal is to develop learning AI systems capable of adapting to new environments while also retaining what they have learned, for life.

Neuroscience and nature continue to inspire the question of building adaptable learning systems.

- ## Psychological Models

Psychologists have conducted numerous studies to determine human performance in certain learning tasks.

Some of these studies focus on reinforcement learning that has some of its work traced back to efforts in modeling the way reward stimuli in animals influence them to learn goal-seeking behavior. This is a vital theme in machine learning.

- ## Evolutionary Models

Nature has it that individual animals not only learn to perform better. The whole species learn or more accurately, *evolve* to be better suited to their individual places.

In computer systems, the distinction between the two can be blurred so that certain characteristics of biological evolution are modeled to techniques to improve the performance of computer programs.

- ## Philosophy

  Occam's learning, also known as Occam's razor, suggests that the simplest hypothesis or the one with the least assumptions is the best one.

  This philosophy also applies to machine learning. By applying Occam's razor, you can enhance a simpler hypothesis to balance between bias and variance.

- ## Artificial Intelligence

  Artificial intelligence and machine learning have been concerned with each other right from the start. Researchers of artificial intelligence have studied and explored the role that analogy plays in learning.

  One of their particular concerns is how you can base future decisions and actions from previous model cases. Some of the methods both AI and machine use include deductive logic programming, decision tree, and explanation-based learning.

- # Theories

  Several theories influence and interconnect with machine learning. To better understand machine learning and how it matters today, knowing the theories underlying the concept is essential.

Among the theories connected with machine learning are adaptive control theory, computation complexity theory, and information theory.

- **Adaptive Control Theory**

Controlling a process with unknown parameters that have to be estimated during an operation is one of the problems that control theorists study.

The parameters usually change during operation, which the control process has to track these changes. Controlling a sensory-based robot has some aspects that share instances of this same problem.

- **Computational Learning Theory**

Computational learning theory, or simply learning theory, is a branch of theoretical computer science. It is dedicated to learning machine learning algorithms' design and analysis.

It deals with the hypothetical limitations of varying learning tasks required to learn, which is measured in computations.

- **Information Theory**

Information theory deals with the movement and transformation of information, which shares concepts and application in machine learning.

An important concept of information theory – entropy, which measures the amount of uncertainty of a random process's outcome – shares similar process in providing predictions after a machine's training sequence.

Chapter 4

Successful Applications of Machine Learning

There is a good reason why machine learning is a catchphrase in the tech world these days. This concept marks a step forward in how we can develop computers capable to learn.

Basically, machines have a machine learning algorithm that has a "teaching set" data.

This data is what allows it to answer questions, provide better decisions or predict the future.

For instance, the teaching set given to a computer is that of photographs that say "this is a boy" for some and "this is a girl" for others.

Once the computer learned that teaching set, you can show it a series of new images.

This time, it will start identifying the photos and separate them to "boy" and "girl". This generates a new teaching set that the machine learning algorithm will add to its knowledge.

The idea is that each photo identified correctly or incorrectly is added to the computer's teaching set.

This effectively makes the machine smarter so that it becomes better at performing and completing its job over time. It is not the same as how humans learn but in a sense, the machine still learns.

1. Speech Recognition Systems

The most successful and effective speech recognition systems use some form of machine learning.

One example of this is the SPHINX, a program that can learn strategies specific to speakers to recognize words and phonemes from the detected speech signal.

Siri and Cortana, which are two of the most common voice recognition systems today, also use machine learning as well deep neural networks.

Machine learning allows them to mimic human interaction and as it learns, they begin to understand the semantics and nuances of human language.

2. Autonomous Vehicles

Machine learning and data also fuels autonomous vehicles or self-driving cars. Machine learning techniques have been used before for training autonomous vehicles to steer appropriately when it drives different types of road.

Top auto executives are also expecting that by 2025, the world will see smart vehicles on the road through machine learning.

These are cars capable of not only integrating into the Internet of Things. At the same time, it learns things about its owner and the environment it to auto adjust.

3. Classification of New Astronomical Structures

NASA also uses machine learning to learn general regularities contained in the data within a variety of huge databases.

For instance, the space agency uses decision tree learning algorithms to find out how to classify the celestial objects in the second Palomar Observatory Sky Survey.

Data mining algorithms have been used to deal with problems such as star-galaxy separation.

Typical surveys have a sheer number of stars and galaxies that separating them needs to be automated.

Mixture modeling, DT, ANN, and SOM are among common data mining algorithms used for this problem.

4. Playing Games Like Backgammon

Machine learning algorithms have also been used in the most successful computer programs used for playing games.

For instance, TD-Gammon is the world's top backgammon computer program that learned its strategy through a more than one million practice games it played against itself.

Through these practice games, TD-Gammon is able to play a competitive level of a human world champion.

Computer games using machine learning also serves as learning ground for other applications such as self-driving systems. Games like *The Open Racing Car* helps train autonomous cars.

5. Personal Security

If you have attended a big public event before or flown on an airplane, you have likely experienced waiting in a long line of security lines for screening.

Now more than ever, providing your identity has become essential especially in the digital landscape.

To cater to this demand, machine learning and artificial intelligence are rapidly created to help in identity authentication.

Machine learning is proving that it can help in eliminating security false alarms. Furthermore, it helps spot things that human screeners might overlook during security screenings.

6. Data Security

Security in the digital landscape is extremely important as malware continues to be a huge and growing problem.

Every day, there are thousands of new malware created. To combat these attacks, machine learning is being widely used in a broad range of security applications.

Facial Check with Video, which was developed by Onfido, is one of the latest security applications using machine learning.

It encourages users to film themselves performing random movements. This can be used for proving the user's identity by checking similar with user's facial image.

7. Healthcare

The use of machine learning allows processing of more data. It can also better detect more patterns than humans.

One example of the advantage of machine learning in healthcare involves a study for computer-assisted diagnosis (CAD).

In the study, the computer is able to detect 52% of the cancers in a review of mammogram scans of women, about a year before they are officially diagnosed. Using machine learning is also ideal for understanding the risk factors for diseases within large populations.

8. Fraud Detection

Machine learning is becoming more and more adept at spotting the possible cases of fraud in a wide array of fields.

It is proving to have potential in sweeping the cyberspace and making a more secure place. It shows many exemplary benefits including tracking of monetary frauds online.

One example of the use of machine learning in fraud detection is the way PayPal uses ML to protect itself from money laundering.

The company does this by using a variety of tools to distinguish legitimate or illegitimate transactions by comparing millions of transactions that take place.

9. Marketing Personalization

Companies also use machine learning in marketing. To better serve their customers and sell well, they need to understand them first.

When you browse an online store but do not buy the product, soon you will see digital ads for the same product while surfing the web.

By using your web search info, companies can personalize that what ads to show you or which email to send to you.

Machine learning and data mining allow them to know you through the websites you go to or links you click. All these information help lead consumers to a sale better than before.

10. Online Search

Online search is probably the most famous applications of machine learning.

Google and all the other search engines constantly improve their search engine standards using machine learning methods. They use a machine learning algorithm to learn from your responses.

Whenever someone executes a search on Google, its program observes how you react to the search results.

If you stayed on the first page, then it assumes you found what you are looking for. If you clicked next, the program assumes it did not deliver the results you wanted.

With that information, it can learn to deliver better results in the future. One of Google's machine learning-based algorithms is Rankbrain.

This was updated to work concurrently with Hummingbird. With the two working together, Google is better able to understand what your queries mean.

11. Recommendations

Surely you are familiar with Netflix, Amazon, and similar services, right? These services also use smart machine learning algorithms to better analyze your activities within their platforms.

Upon analyzing and learning about them, ML algorithm compares it with millions of other users.

This will help online platforms like Netflix to determine what other services or products you might like. These recommendations get better every time you use the platform.

Any new data about you will be incorporated and learned by the algorithm to refine its recommendations.

12. Financial Trading

Many people out there want to be able to predict what will happen to the stock market anytime and any day for apparent reasons.

One way to predict this is the use of machine learning algorithms, which are getting better and closer at cracking the financial trading all the time.

Today, many trading firms are using machine learning systems able to predict and perform trades at high volumes and high speeds.

Most of these ML systems depend on probabilities that can turn a trade, with significantly low probability but has high enough speed and volume, into huge profits.

13. Natural Language Processing (NLP)

Natural language processing or NLP is being used in varying kinds of interesting applications across disciplines. Together with machine learning, it found great use in the interactions industry as a stand-in for customer service agents.

Using machine learning together with natural language, customer service agents can better and more quickly direct customers to their needed information.

One example of machine learning and NLP is a chatbot that is capable of translating and interpreting human natural language input.

14. Email Spam and Malware Filtering

Email clients use a wide range of spam filtering methods to segregate email spams. To make sure that these spam filters are updated continuously, the filters are powered by machine learning. Typically, rule-based spam filtering is unable to track the latest spam tricks by spammers.

However, machine learning like C 4.5 Decision Tree Induction and Multi-Layer Perception are two of the ML-powered techniques you can use for spam filtering. Besides spam filtering, ML is also great in detecting malware.

Machine learning-powered system security programs can better understand malware' coding patterns.

Hence, these security programs are better able to spot new malware showing 2 to 20% variation in their coding patterns. ML-powered security systems are able to offer better protection.

15.   Traffic Predictions

We are commonly using GPS navigation services now to locate and get better directions to our destination.

When we use GPS, our current locations, as well as velocities, are saved at a central server, which is used for managing traffic.

This information is used for constructing a map of the present traffic. There is an underlying problem – the number of cars equipped with GPS is considerably less.

Using machine learning in traffic and similar scenarios, it becomes easier to evaluate the areas where there's congestion every day.

In addition to the examples mentioned above, machine learning is proving its potential in more ways. Many people use them and experience ML changing their day-to-day life.

Now that you know more about machine learning, you can probably see it in different aspects of your life.

Chapter 5

Important Terms and Processes in Machine Learning

Algorithm

The machine learning algorithm is the set of hypothesis taken at the start before training using real-world data starts. In mathematics and computer science, it refers to an explicit specification of the method to solve a class of problems or tasks.

Algorithms can perform calculation, automated reasoning, and data processing tasks. You can divide machine learning algorithms into three different categories. Each of them having broad scope and wide array of algorithms you can use for different kinds of problems.

The first includes supervised learning, meant for cases where there is a label for the training set used for predicting the correct response or decision under the current circumstance.

Unsupervised learning, on the other hand, is practical in such cases where the task is to determine the given unlabeled dataset's implied relationships.

As for reinforcement learning, there are no labels, but it is accompanied by positive or negative feedback depending on the algorithm's proposed solution.

Model

In machine learning, a real-world process can be represented in mathematical expression through a machine learning model. Creating a machine learning model needs you to provide training data from which the machine learning algorithm can learn and make predictions or decisions.

A mathematical model uses mathematical language and concepts to describe a system. It is used in several engineering disciplines such as artificial intelligence, computer science, and machine learning. Using a model helps in explaining a system, studying its components' effects and make predictions.

Three of the most common types of machine learning models include binary classification, multi-classification, and regression. The type of model you use for your learning task depends on what kind of target you are aiming to predict.

- *Binary Classification Models* – Models used for binary classification predicts a binary outcome, meaning one of two possible classes.

- *Multi-Classification Models* – Models used for multi-classification permits generating of predictions for multiple classes, meaning that you predict one out of two or more outcomes.

- *Regression Models* – Regression models are meant for regression problems and used in predicting numeric value.

Input Vectors

Input vectors, or feature vector, pattern vector, sample, example, or instance, the input data are known by many names.

On the other hand, its components are also known as attributes, features, components, or input variables. There are three main types of the components' values – discrete-valued numbers, real-valued numbers, or categorical values.

An example of categorical values is student information, which has attributes with values represented by sex, major, adviser, class, etc.

For instance, a particular student may have a vector represented as [female, education, Smith, junior]. Another quality of the components is that the categorical values can either be ordered or unordered.

Similarly, it is also possible for all these types to come in mixtures.

In any case, it is also possible for the input to be represented in an unordered method by listing the attributes' names and the values together. The attributes are assumed to be in order by the vector form as given tacitly.

An example of an attribute-value representation is: [Sex: Male; Age: 20; Class; Junior; Adviser: Smith]. There is also another specialization in representing input data, which uses Boolean values. This specialization can be viewed as a special case where it is either categorical variables [*Yes, No*] or discrete numbers [1, 0].

Outputs

In machine learning, the output can be a real number. In this case, the process of representing the function h is known as *function estimator*. On the other hand, the output is referred to as *an estimate* or *output value*. The output may also be of categorical value.

In this case, the processing of representing the function h is called in several names – a *recognizer, classifier,* or *categorizer*.

As for the output, it is also referred to using several names – a *class, label, decision, or category.* Classifiers find use in several recognition problems, such as in the classification of handwritten characters.

In this learning problem, the input is an appropriate embodiment of a printed character. The classifier then maps the input into one of several different categories. Additionally, the components of vector-valued outputs may also be categorical values or real numbers.

There is also an essential special case when it comes to outputs – the Boolean output values. If this is the instance, the training pattern with a value of 1 is referred to as a *positive instance* while the training pattern with a value of 0 is referred to as *a negative instance*.

If you have a Boolean input, the classifier fulfills a *Boolean function*. If it is a Boolean case, you usually have to study it meticulously because this instance allows making general points within a basic setting. Also called *concept learning*, the function of a Boolean function is referred to as a *concept*.

Training Set

In machine learning, the training set is the dataset used for training the model. To train the model, you pick out specific features from the training set. These features will then be assimilated into the model. If labeled correctly, the model should learn something from the features obtained from the training set.

Incremental Method

You can use the training set to yield a hypothesis function in a number of ways. One method is the *incremental* method where you select at a time one member of the training set and use this alone to adjust the existing hypothesis at the time.

Afterward, you select another member of the training set, so on, and so forth.

You can have the member selected randomly with replacement or you can repetitively cycle through the training set. If the training set only becomes available in a single member at a time, you can use the incremental method. In this method, you select and use the training set member whenever it arrives.

Online Method

When you use the members of the training set when they become available, it is referred to as an *online* method. This type of method is typically used when, for example, the following training instance is a certain function of the present hypothesis as well as the previous instance.

In this case, you used the classifier to decide on a machine's next action using the set of the current sensory inputs. The next set of sensory input will be based on the selected action. In this method, the weight parameters constantly update so the error calculation uses altered weight parameters.

Batch Method

Another method is the *batch* method where you have the entire training set available and used simultaneously in calculating the function h. This method also has a variable where the entire training set is used to repetitively change the current hypothesis until you obtain an acceptable hypothesis.

In *batching*, you provide a portion of the dataset during every repetition, the part where you compute loss or gradients, and where backpropagation occurs. When you say batching, what you actually mean is you are using 100% of the dataset.

Noise

There are times when the training set's vectors become corrupted by noise. An example of noise is when you have sensors and the values you received change even if the recorded signal did not change. The error in the model is simplified as noise.

The noise in observed data can be anything. In an image classification problem, your task is to give labels to images. In this problem, the model errors created by the human labelers can be called as noise.

In that sense, noise is undesirable effects described with random variables because you cannot describe a deterministic model in a sufficiently simple way.

Noises come in two kinds – class noise and attribute *noise*. Class noise is a kind of noise that randomly changes the function's value. On the other hand, attribute noise randomly changes the values of the input vector's components. In either case, you cannot assert that the hypothesized function agree accurately with the values in the training set's samples.

Performance Testing

While inductive learning does not have any correct answer, methods for testing or evaluating the learning's result remains essential.

In supervised learning, a *testing set* is used to evaluate the induced function. The testing set is a discrete set of function values and inputs.

If the hypothesized function is determined to predict well on the testing set, then that means it *generalize*. In machine learning, the acceptance criteria are not expressed in terms of type, amount or severity of the defect. In most cases, the acceptance criteria are expressed in statistical likelihood of approaching within a particular range.

Features

The discrete independent variables acting as the system's input are the features. Features are used by prediction models in making predictions.

Using old features, you can get new features using the method called "feature engineering." Features are also called attributes, components or input variables while the number of features is referred to as dimensions.

Label

Labels are the task's final output, such as the output classes. When the researcher or data scientist said labeled data, they indicate groups of samples that have been marked with one or more labels. In a classification problem, say for example an email filtering learning problem, emails tagged as spam or non-spam are labeled data.

Training

When a machine undergoes training when it passes training data. During training, the learning algorithm determines patterns within the training data, for instance, the input parameters agree with the target.

After the training process, the output is a machine learning model that you can now use for making predictions. This process is also referred to as "learning."

Target

The input variables' output is the target. In a classification problem, the target could be the individual classes mapped to by the input variables.

In a regression problem, the target could be the output value. If it is a measured training set, then the training output values to be measured is the target for the learning problem.

Overfitting

In machine learning, how well the target function's approximation has been trained with the use of the training data to generalize new data is an important consideration. If the training data's sample has a high signal when it comes to noise radio, it is said that generalization then works best.

Otherwise, you will have poor generalization and poor predictions. It is said that if the training data fits too well and new data is poorly generalized, then you have an overfitting model.

Chapter 6

How Machine Learning Works

How does machine learning work? There is a common principle underlying all machine learning algorithms, supervised or unsupervised, intended for predictive modeling.

In this chapter, you can learn about the common principles that all algorithms share so you can discover how they work and better understand the idea behind machine learning.

Learning How Input-Output Function Work

In machine learning, there is a function f and your task is to find out what that f is. You use h to denote your hypothesis regarding the function. The input data $X = (x_1, x_2, \ldots x_i, \ldots, x_n)$, which has vector value and n components have f and h as functions.

A machine with X input and $h(X)$ output implements h with both f and h possibly vector-valued. The assumption *a priori* is that you selected the hypothesized function h from a class of functions H. There are times when you know for certain that f belongs to H or to its subset.

A training set m of input vector samples are used to select the h. The assumptions you made concerning all these entities relate to a number of important details about how a machine learning algorithm works. To further delve into this, consider the two settings in which learning can happen.

The first setting is *supervised learning* where the training set's m samples can have known or roughly known values of function f. If the training set is large, then this learning setting can have a hypothesis that has a good prediction for f.

The other setting is *unsupervised learning* where the vectors' training set does not have corresponding function values.

In unsupervised learning, the problem is usually to appropriately divide the training set into subsets.

This problem can still be regarded as one where you learn a function. In this setting, the function's value is the same subset the input vector belongs.

Predictive Modeling

Machine learning's most common type is one where you learn the mapping of $Y = f(X)$ to make predictions of Y for new input X.

This type of machine learning is referred to as predictive modeling, also known as predictive analysis. The goal of the learning task is to make the most precise predictions possible.

That being the case, you are not really interested in finding the form or shape of the function f you are learning. The only thing you truly care about is making accurate predictions.

To learn the relationship between data, you need to study the mapping of Y = f(X), which is known as statistical inference.

If you are learning function f, that means you are approximating its form based on available data.

If that is the case, your estimate will likely have an error. It will not make a perfect approximation for the fundamental proposed best mapping from output variable Y given input X.

In the application of machine learning, a lot of the time spent is in trying to improve the underlying function's estimation.

An improved estimation will result in the model's predictions to have improved performance. Machine learning techniques are used to calculate approximately the target function.

Learning Requires Bias

The learning algorithm's inductive bias, also known as learning bias, is the set of assumptions used by the learner in predicting outputs with unknown set inputs.

In machine learning, the goal is to develop an algorithm capable of learning to predict a particular target output.

To accomplish this, training examples that establish the relationship between input and output values that you envisioned has to be presented to the learning algorithm. Supposedly, the leaner estimates the correct output for new examples that were not shown during training.

You cannot solve the problem exactly if there are no additional assumptions. This is because there might be random output value to unseen situations that could influence the algorithm's predictions. Inductive bias refers to the type of necessary assumptions about the target function's nature.

Occam's razor is a typical example of inductive bias. Occam's razor assumes that the simplest and most consistent hypothesis of a target function is the best hypothesis.

A consistent hypothesis means the learner's hypothesis produced correct outputs for the entire examples given to the algorithm.

Learning Needs Bias Training to Overcome Stereotypes

Bias is necessary for machine learning but there it also has dangers that must be blocked. It can be useful, say for example when you decide not to step in front of an accelerating bus.

Most, if not all, people are biased towards that. However, it is not a good thing when it unfairly disadvantages one over another.

Unconscious biases in humans are deep-seated assumptions about certain personal attributes, which can influence a decision maker's decision making without being obviously aware. These biases are collective "mental shortcuts" as they say that people make based on stereotypes and social norms.

It said that unconscious biases have an important influence on one's decisions.

For instance, in the business setting the hiring manager's unconscious bias can influence his or her decision to hire an applicant. On that note, many businesses take measures to understand and stop it.

However, even though people's biases can be taken care of, the systems they build seem to be vulnerable to it. Take for example personalized online marketing.

Experiments on Google ads found there is significantly less number of online ads shown to women than men, which promises to help them get a job.

This shows a clear gender bias, which raised some questions regarding the fairness of the targeted online ads. Besides gender bias, another issue is a racial bias that was uncovered based on Google searches.

It was uncovered that names that identify black people produced higher incidents of "arrest" related ads than with names that identify white people.

In both cases, no programmer wrote any algorithm explicitly racist or sexist. The true culprit that made the biases were the machine learning algorithms, which automatically learn patterns based on the massive data sets presented to them. Just like humans, machine learning algorithms can also develop biases.

These biases, if not checked and corrected can lead to an algorithm to develop discriminatory behavior. Learning algorithms develop biases due to several possible reasons.

One way they can learn bias is through the training data's selection-bias. This happens if the model was trained using a dataset that does not embody the population.

If that is the case, then it will likely make faulty general inferences. Another reason that a learning algorithm can learn bias is through hidden variables.

One possible way to prevent biased machine learning algorithms is ensuring you do not include data that might cause such problems. For instance, you can remove gender and race data, which often causes bias.

Now that machine learning is expanding to even sensitive areas such as in hiring and credit scoring, it goes to say keeping algorithms unbiased is essential.

The first thing that can be done to achieve this goal is by raising awareness about the existence of biases in machines, particularly the serious negative impact it might have.

It is important that machine learning models are tested for discriminatory biases. The companies that created those learning models should publish the results of their test and share the datasets as well as useful methods used in those tests. Then such tests will be publicly reviewed to make the process more effective and easier.

What is a Well-Define Learning Problem?

To better understand machine learning, it is essential to consider a few learning tasks. But before that, let us define learning broadly to be any computer program capable of improving its performance at a task through experience.

As an example, let us consider a computer program that can learn to play checkers.

The computer program can improve its performance, which is measured by being able to win the class of tasks associated with playing the game of checkers. How it improved is through the experience it gained by playing checker games against itself.

If you recognize, a well-defined learning problem has three features – the training experience E, a measure of performance P, and the class of tasks T.

In the checkers learning problem above, all of these three are present. The task is to play checkers and the training experience involves it playing checkers against itself.

On the other hand, how many times it won against opponents serve as a measure of performance. These three are vital for easily specifying the learning problem, such as with learning a self-driving car and a database of handwritten words and letters. All these learning problems have three things in common.

Checkers Learning Task

- **T** – Playing checkers.

- **P** – The percentage or number of games it won against challengers.

- **E** – The computer program practicing by playing games counter to itself.

Self-Driving Car Learning Task

- **T** – Driving in public lanes while using vision sensors.

- **P** – The average distance the car traveled before a human overseer judged that there is an error.

- **E** – A series of images and command the car recorded while it observed the human driver.

Handwriting recognition Task

- **T** – Distinguishing and categorizing handwritten words.

- **P** – The percentage of words the machine correctly classified.

- **E** - The machine is given a database of handwritten words already classified.

Speech Recognition Task

- *T* – Recognizing speech patterns.

- *P* – The percentage of phonemes and words correctly recognized.

- *E* – A series of spoken words and sentences recorded or spoken by the user.

Filtering Spam Email Task

- *T* – Recognizing and classifying emails into a spam or non-spam.

- *P* – The number of emails correctly classified into spam or non-spam class labels.

- *E* – A database of emails with given classification.

Chapter 7

Machine Learning Techniques

Both big data and cloud computing are becoming more and more important today because of their contributions.

As they gain more importance, machine learning is the technology that can analyze the big pieces of data that these two provides.

This way, the automated process reduces the burden of the tasks of data scientists.

Most of the techniques used for data mining have actually been around for many years.

However, these techniques were not as effective because they did not have the competitive power needed for running algorithms.

If you have access to better data and you run deep learning, it output leads to significant breakthroughs, also known as machine learning.

Figure 1. Machine Learning Techniques/Types of Machine Learning

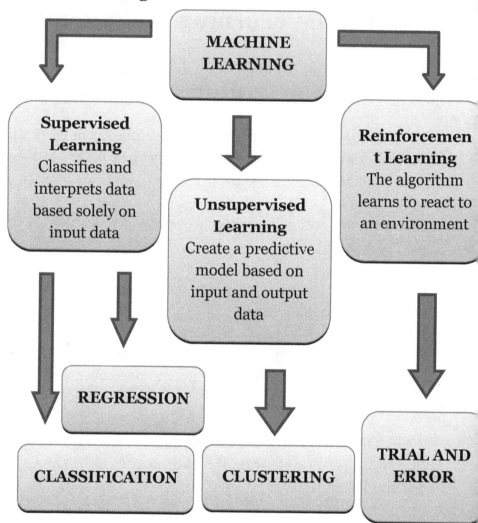

Three Types of Machine Learning

There are different types of machine learning based on the algorithm and its purpose. The two main types/ techniques used for machine learning are *supervised learning* and

74

unsupervised learning. The former builds a model capable of making predictions based on known data in the presence of uncertainty.

The latter, on the other hand, deals with finding hidden patterns or intrinsic structures in input data.

Besides these two, there is another type of machine learning that is not more familiar – *reinforcement learning.* This is really powerful machine learning that is also complicated to apply for learning.

Supervised Learning

Supervised learning happens when a machine learning algorithm learns from data taken as an input, which is called *training data.*

Besides the training data, it can also learn from related target responses that may include string labels or numeric values.

For example, tags or classes that will later help in predicting the correct response when new examples are presented. Most of applied machine learning utilizes supervised learning. Its main goal is to make predictions by learning a model from labeled training data.

In this type of machine learning, the term *supervised* refer to the set of samples that have known desired labels and output signals *(targets).* Basically, the system or structure attempts to learn after the given previous examples.

This is in contrast to unsupervised learning where the system tries to find the forms, patterns or arrays directly from the given example.

Putting it in mathematical perspective, supervised learning has both input variables and output variables. You can obtain the mapping function from the input to the output using an algorithm.

If the input is X and the output is Y, the mapping function will be written as $Y = f(X)$.

Let us consider algorithm for filtering email spams as an example. You can use a supervised learning algorithm on a mass of categorized emails. These are emails that have correct labels, whether they are spam or none spam.

Using the algorithm, you can train a model to predict if a new email goes to the spam category or none spam category. These are two subtypes of supervised learning – classification and regression.

Figure 2. *Supervised Learning*

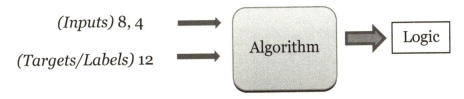

(Inputs) 8, 4

(Targets/Labels) 12

Algorithm

Logic

Training with data

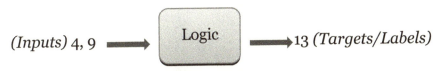

(Inputs) 4, 9

Logic

13 *(Targets/Labels)*

Predicting with New Data

Included in the training data are both the inputs and labels or targets. The figure above is a simple demonstration of the process of supervised learning.

The first figure involves the training, where the idea is for the algorithm to learn how to add two numbers, a=8 and b=4, which are the inputs.

It also includes the target or label, 11, which is the result when the two numbers (input data) are added. After practicing and learning with the help of the training data, the algorithm learns logic. That is, it learns how the addition process is done.

With that information, the algorithm has now learned how to predict (or add, in this case) with new information. As shown in the next figure, the new data includes two new inputs (4, 9). With its new knowledge, the algorithm is able to answer it correctly (13) as the target or label.

Image 1. (a) Classification and (b) Regression

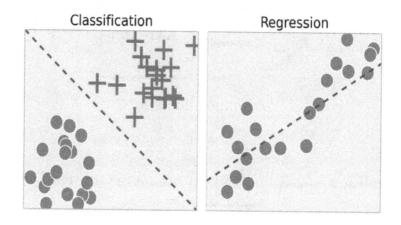

Classification   Regression

Classification

It is a classification task or problem if the output variable is a group or a category, such as an email either a "spam" or "non-spam".

Classification is a subtype under supervised learning. The goal is to predict the class labels of new occurrences based on previous observations. The class labels are unordered and distinct values, which can be understood as associated with the occurrences.

- **Binary Classification**

    One classic example of a binary classification task is the email spam detection mentioned before. In this example, a set of rules is learned by the machine learning algorithm.

These rules are set to differentiate the emails between two possible categorical classes – spam or non-spam. But the categorical class labels do not have to be binary.

When a supervised learning algorithm learns a predictive model, it can designate any class label from the training data set to new occurrences that are yet to be labeled.

Handwritten character recognition is a classic model of a multiclass classification task. In this task, you would gather a training dataset comprised of multiple examples of handwritten copies of every letter of the alphabet.

This way, if a new handwritten letter is provided by the user through an input device, the predictive model can predict the right letter with a certain degree of accuracy.

If the character provided is any from the digits zero to one is, the machine learning system might not be able to predict or recognize it accurately. This happens if the new character is not part of the training dataset.

For example, let us take a binary classification task given 20 training samples. In these training samples, 10 are categorized as the positive class (indicated by the plus signs), while the other 10 are categorized as the negative class (indicated by the minus signs).

This scenario makes a two-dimensional dataset.

This means that every sample in the dataset is associated with two values. We can write it as X_1 and X_2.

To separate the two classes, the supervised machine learning algorithm learns a rule that will classify new data into each of the two categories. When new data is provided, the machine learning algorithm will separate it either as an X_1 or X_2 value.

- **Multi-Classification**

  A subtype in the classification technique in supervised learning, multiclass or multi-classification is the task of classifying cases into one of three or more classifications.

  Some classification algorithms allow more than two classes to be used, most are binary algorithms in nature. However, even though the algorithms are inherently binary, it is possible to turn them to multiclass.

  Multi-class classifications problems often arise in several application domains such as computer vision, biology, information retrieval, and social network analysis where data occurrences do not belong singly to a certain class.

  Some obvious examples of multi-level classification problem are text classification and image classification where they can have multiple topics.

Regression

The regression type of supervised learning aims to calculate a continuous value. This type of learning permits you to approximate a value, for instance, a human lifespan or the housing prices, from input data. In this, *target variable* refers to the unidentified variable that you aim to predict.

Continuous target variable means is that the value of Y cannot take on discontinuities or gaps. Examples of continuous values are the height and weight of a person.

Another type of variable in regression is *discrete* variables, which means it can take on values of a limited number. An example of this is how many children a person has.

A typical regression problem involves predicting income. In this problem, all the pertinent data about the individual, such as job title, years of experience, years of education, etc., corresponds to input data within the dataset.

You can use this information to predict the person's income. Called *features*, the attributes of the information can be categorical or numerical.

In order to find the target output Y, you will want to have numerous training observations relating to the input data X's features as you possibly can.

The more training you have, the more that the model is able to learn of the relationship or function f that exists in X and Y.

There are two sets of data – the training data set and the test data set.

The table below could show you a trivial look at a simple 2D training data containing a row that includes the level of education and income of a person.

You can also add new columns to contain more features of the input data. This will make a more intricate but also possibly a more precise model. Take a look at the example table below.

Table 1. Supervised Learning: Regression

Observation No.	Education Level (No. of Years in Higher Edu)	Yearly Income
1	5	$96, 000
2	2	$45, 000
3	6	$105, 000
4	0	$39, 500

Training Set

To make accurate and practical predictions about a person's income (output Y) based on his or her education level (input X), you need to build the right model. You can do this through the use of algorithms for supervised learning, which you can familiarize yourself more in the next chapter.

Unsupervised Learning

In an unsupervised learning, you find intrinsic structures or hidden patterns within data. In this type of machine learning, you draw inferences from the dataset that contain input data with unlabeled responses.

This is a machine learning technique that is quite useful when you are not exactly sure what you are looking for. This type of learning happens when the algorithm learns from ordinary examples with an unassociated response. In this case, the algorithm is left to find out the data patterns on all on its own.

In unsupervised learning, the data is often restructured by the algorithm into something else like a new series of unassociated values or new features representing a class.

The restructured data are great and practical in providing insights into a data's meaning. It also provides supervised machine learning algorithm with new valuable inputs.

This kind of learning is somewhat alike with the methods humans use in determining if two events or objects are from the same category. Unsupervised learning is much the same as when humans observe how similar two objects are.

This is the same type of learning that some recommendation systems on the web, which use marketing automaton algorithm, are based. The suggestions of this recommendation algorithm are derived from those you have bought in the past.

The algorithm is able to make recommendations based on an estimate of what group of customers an individual resembles the most. From that knowledge, it is able to infer what your likely preferences are, which is based on the group's common preferences.

Compared to supervised learning, coming up with metrics to determine how well an unsupervised learning algorithm does is not always easy. In this type of learning, performance is usually subjective and specific to a domain.

Figure 3. Unsupervised Learning Process

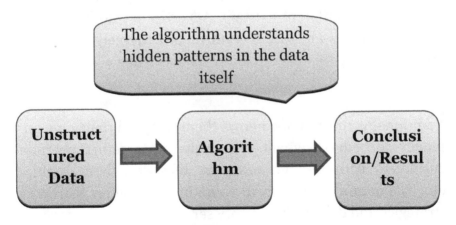

Clustering

The most common technique for unsupervised learning is clustering. This machine learning technique uses exploratory data analysis to determine hidden groupings or patterns in data. Some of the common applications of clustering are market research, gene sequence analysis, and object recognition.

Personicx, a life stage clustering system created by Acxiom, a marketing data provider, is a fascinating example of clustering.

This system classifies U.S. households segmenting them into 70 discrete clusters in 21 life stage groups. Advertisers use this data when targeting direct mail campaigns, display ads, Facebook ads, and more.

According to Personicx's white paper, they use principal component analysis and centroid clustering in putting U.S. households in different clusters. Advertisers who have access to these clusters are sure to find them extremely useful. For one, it will help them get to know their existing customer base better.

Additionally, this method allows them to spend their time and money for advertisement effectively by targeting potential customers with relevant interests, lifestyles, demographics, etc. The data allows them to come up with better marketing strategies and deliver satisfactory service to their customers.

Reinforcement Learning

Reinforcement learning happens when the algorithm is presented with unlabeled examples, just like in unsupervised learning. However, the example can be accompanied by positive or negative feedback based on what the algorithm's proposed solution is.

In this type of learning the product is not just descriptive but also prescriptive. This is because it is connected to the applications that it has to make decisions for and bear its consequences. To put it simply, it is just like *trial by error*.

Learning by trial and error helps you learn because the mistakes have an added penalty for cost and pain. This added penalty teaches you that you are less likely to succeed if you follow a certain type of action. One example of this learning is a computer to learn to play video games on its own.

It is likely that in this case, the algorithm is presented with examples of exact situations. For instance, it showed situations of the gamer being stuck inside a maze while trying to avoid an enemy. The algorithm knows of the outcomes of the actions it takes through the application.

It learns while it tries to avoid the situations it discovers to be dangerous while it pursues survival. The difference between supervised learning and reinforcement learning is there is always an answer key for the "supervisor" together with the training data.

In reinforcement learning, the learning agent needs to decide the right action in performing its task even though there is no answer key. If there is no training data, it learns instead through experience. Instead of training data, it gathers training examples.

Training examples are its experiences of what action is good and what action is bad while performing the task. The algorithm learns while attempting the task through trial and error. Its goal is to maximize long-term reward. As an example, to learn reinforcement learning, consider a game of a robot mouse inside a maze.

The game you are playing includes a mouse inside a maze. Using a point system, the goal is for the mouse to get to the final reward of a huge cheese that will earn +10000 points, or the lesser reward of water for +100 points.

The mouse also wants to avoid the places delivering electric shocks (-100). While exploring, the mouse might find the lesser reward of water sources just near the entrance.

After finding these, it might spend all its time there exploiting its discovery. As a result, it never goes further inside the maze to find the much better and larger prizes. What happens then is it misses out on the ultimate reward waiting inside the maze. This is what is called the *exploration/exploitation* tradeoff.

A simple exploration study is for the mouse to take the best course of action most of the time, say about 80%. At the same time, it explores new direction randomly selected sometimes. That is even though it means the mouse will walk away from the known reward. This is referred to as the epsilon-greedy strategy.

In this technique, the epsilon refers to the time the agent takes a randomly selected action. This is in contrast to the percent that the agent takes action that will likely maximize the given reward given what is known.

In this learning, you typically start with plenty of exploration, which means higher epsilon value. As time goes by, the robot mouse learns more regarding the maze.

During that time, it also learns what actions will produce the most long-term benefits. With that, the robot mouse begins using what it knows, which leads to lower epsilon value.

In machine learning, the reward does not always come immediately. Often, the reward comes after a long stretch of time after the learner has gone through several actions and several decisions.

In this process, the learner has to observe its environment then decides to interact with it where it receives a positive or negative reward.

There are many practical applications to reinforcement learning as a machine learning technique. Some of these applications are in the industry of manufacturing, delivery management, inventory management, and more.

Figure 4. *Reinforcement Learning Process*

Chapter 8

Approaches to Machine Learning

There are so many algorithms available that when you hear their names, it is easy to be overwhelmed. You feel like you are expected to know what each of them is and what kind of problem to use them for.

Getting yourself familiarized at least with the most popular of them is the key not become overwhelmed. On that note, this last chapter of the book deals with that.

How to Decide Which Machine Algorithm to Use

As mentioned before, there are dozens of machine learning algorithms out there that choosing the right one seems overwhelming. Each of these methods follows a different approach to learning. There is no best method and not one that fits all problems. Finding the right algorithm to use is somewhat trial and error.

Sometimes, even experienced data scientists cannot tell if an algorithm works without trying it first. However, the selection of the algorithm to use is also based on your data, e.g. its size and type. Furthermore, the kind of insights you are looking for in the data and how these insights will be used influences the algorithm selected.

Classification Techniques

These techniques predict separate responses, such as if an email is a spam or not, or if a tumor is benign or cancerous. Models using classification techniques classify the input data into different categories.

These techniques are typically applied in credit scoring, medical imaging and speech recognition. Some of the most common classification algorithms are:

- *K*-Nearest Neighbor
- Support Vector Machine
- Naïve Bayes
- Logistic Regression
- Neural Networks

Regression Techniques

These techniques' purpose is to predict incessant responses, such as fluctuations in the demand for power, or temperature changes.

Regression techniques are often used if your response's nature is that of a real number or if you are operating using a data range. Some of the most popular regression techniques are:

- Linear Model
- Non-Linear Model
- Stepwise Regression
- Neural Networks

- Bagged Decision Trees

Clustering Technique

Clustering is the typical technique for unsupervised learning. You use this technique for analysis of experimental data with the aim of finding hidden groupings or patterns.

Some of the common applications of clustering technique are market research and gene sequencing analysis. Some of the algorithms that use clustering technique include:

- Hierarchical Clustering
- K-Means
- K-Medoids
- Subtractive Clustering
- Hidden Markov Models

Common Machine Learning Algorithms

As mentioned many times before, machine learning algorithms can be broadly classified into three. Each of these is ideal for specific real-world problems allowing your job to be simplified and automated. In this section, you will learn the most common algorithms you can use to predict outcomes, make correct decisions, and improve tasks.

Decision Tree

In decision tree learning, the predictive model used is a decision tree that maps an item's observations to conclusions about its target value. The central learning approach involves dividing the training data repetitively into scores of identical members using the most discerning dividing criteria.

The output labels determine how uniform or homogenous the buckets of data members are. If the output labels are of numeric value, the member scores' dimension becomes the variance. If the output labels are categorical, you use Gini index or entropy for the bucket.

Through the learning process, you will try several dividing criteria, which depends on the input. If it is a categorical input, such as Monday, Tuesday, Wednesday, etc., turn it into binary to assess similarity using true/false for decision boundary.

For an input with an ordinal or numeric value, use the 'less Than and greater Than' as decision boundary for an input value of every training data. If you don't find substantial gain after further splitting the Decision Tree, then the process stops.

Prediction is voted by the score members embodied at the leaf node. When it is a categorical output, majority wins. If it is a numeric output, the average wins.

The advantage of the Decision Tree is its flexibility when it comes to the type of data of the input and the output variables.

In this approach, the data type can be binary, numeric, and categorical value. Different values' level of influences is indicated by the degree of the decision nodes. Every limitation of the decision boundary at every split point is a firm binary decision.

Only a single input attribute at one time is considered by the decision criteria unlike with other algorithms where it is possible to consider a group of many input variables. This is one of the weaknesses of this machine learning approach.

Another of its weaknesses is that it cannot be updated incrementally after it has learned. When there is new training data, you will have to discard the old tree then reorient every new data. There are, however, methods to address these limitations.

Nearest Neighbor

The idea with the Nearest Neighbor method is you are not learning a model but instead to search the training set for K similar data point. This data point will be used for interposing the output value. This can either be the numeric output's weighted average or the categorical output's majority value.

In order to choose the best value, you need to cross-validate K, which is a tunable parameter. This algorithm needs the distance function's classification that is used in finding the nearest neighbor. The typical practice used for numeric input is by normalizing them.

You can do that by subtracting the mean, which you then divide using the standard deviation. If you have an independent input, you can use the Euclidean distance. If not, you should use the Mahalanobis distance instead, which accounts for the relationship between the sets of input features.

Jaccard distance is used for binary attributes. The advantage of the K nearest neighbor approach is its straightforwardness. This is because you do not need a model to train using this method. When the data arrives, incremental learning commences as it is automatic.

If you want, you can also delete the old data. Take note, however, that you must organize data using a distance-aware tree. There are limitations to this algorithm as well. For one, it is not capable of handling a huge number of dimensions.

Furthermore, you need to hand tune the increment of the different factors through cross-validation on the various weighting combination. Having to do this is quite a dull process.

Linear Regression

In these methods, the basic assumption is you can express the outside variable (numeric value) as a linear combination or weighted sum of the input variable set (numerical value). This is an example of a linear combination of the input variable set and outside variable: $Y = w1xw + w2x2 + w3x3 \ldots$

During the training phase, the goal is to learn the weights ($w1$, $w2$, $w3$ ...) through minimization of the error function lost (y, $w1xw + w2x2 + \ldots$). One of the typical ways to solve this problem is the *gradient descent* with the concept of altering the weights alongside the direction of the loss function's maximum gradient.

In linear regression method, it is necessary that the input variable is numeric. The binary variable will be represented in 1, 0. Each possible value in the categorical value will also be represented by a discrete binary variable. If the output is a binary variable, you will use the logit function in transforming the −infinity to +infinity range to 0 and 1.

Referred to as logistic regression, you will have to use a different loss function that is centered on maximum likelihood. The method used to prevent overfitting is called the regularization technique where you penalized large value of weights (w_1, w_2 ...). You determine the L1 by adding into the loss function w_1's absolute value.

For L2, you add into the loss function the square of w_1. It is a property of L1 to penalize irrelevant or redundant features more, which makes it a great feature for selection of features that are exceedingly influential. One of the strengths of using Linear Model is that learning and scoring both have high performance.

Similarly, the *Stochastic Gradient Descent-based Algorithm* is proven capable of handling incremental learning. Its weakness is that the input features' linear assumption that is usually false. Thus, it is necessary that every input feature is transformed. You can also use various transformation functions to find one the output can have a linear relationship with.

Neural Network

This machine learning approach is often considered as layers upon layers of perceptrons. Every perceptron is a logistic regression unit that has several binary inputs and a single binary output. When written in mathematical expression, the multiple layers are expressed as: $z = \text{logit}(v_1y_1 + v_2y_2 + \ldots)$, with $y_1 = \text{logit}(w_{11}x_1 + w_{12}x_2 + \ldots)$.

The multiple layer mode allows learning of input X an output Z's non-linear relationship. Known as the 'backward error propagation", this technique has the error propagating from the back of the output layer towards the input layer so that the weight will be adjusted. Take note that Neural Network needs binary input.

This means that you need to first have the categorical input converted to multiple binary variables. You can change the numeric input variable into 101010 string of binary codes. You can also transform the numeric and categorical output using this same method. Without transforming the values into numeric, you cannot use the Neural Network approach.

Bayesian Network

Bayesian Network is much like a dependency graph. In this method, every node embodies a binary variable while every directional edge embodies the relationship's dependency. If, for instance, Node A and Node B take an edge to Node C, it means the probability of C is true.

However, it depends on the varying combinations of A and B's Boolean values. In this approach, learning pertains to finding all the incoming edges' joint-probability at every node. You can do this by calculating A, B, and C's observed values.

After that, you can update NodeC's joint probability distribution table. When each node finally has the probability distribution table, you can begin calculating any hidden node or output variable's probability. You can do the computation from the observed nodes through the Bayes rule.

The advantage of Bayesian network method is the fact that it is highly accessible. Using this method, learning can be done incrementally since all that you really have to do is add all the observed variables then fill in the probability distribution table.

One thing to note is that all data must be binary in a Bayesian network, just like with Neural Network. In that case, you also have to transform the categorical variable so it becomes multiple binary variables. Generally, a numeric variable is not ideal for a Bayesian network.

Support Vector Machine

This method takes in both numeric and binary input. The basis for the approach is on searching a linear plane that maximum margin for separating two classes of output. Just like with Neural Network, you can transform the categorical input into a numeric input.

The categorical output, on the other hand, can be shown as multiple binary outputs. You can also use Support Vector Machine to do a regression using a different lost function. The main advantage of this approach is its capability of handling a massive number of dimensions and non-linear relationships.

These are among the most popular and commonly used machine learning algorithms. If you are considering of entering a career in machine learning, it is best that you start right away. This field is actively growing faster every day.

In that case, the sooner you can understand its scope, then the better you can create solutions that can solve the world's most complex problems. As you have already learned in the previous, machine learning will only grow more relevant in the future.

Final Words

I would like to thank you for purchasing my book and I hope I have been able to help you and educate you on something new.

If you have enjoyed this book and would like to share your positive thoughts, could you please take 30 seconds of your time to go back and give me a review on my Amazon book page.

I greatly appreciate seeing these reviews because it helps me share my hard work.

You can leave me a review on Amazon.com.

Again, thank you and I wish you all the best!